GRAPHIC CAREERS
ASTRONAUTS

by David West

illustrated by Jim Robins

New York

Published in 2008 by The Rosen Publishing Group, Inc.
29 East 21st Street, New York, NY 10010

Copyright © 2008 David West Books

First edition, 2008

Designed and produced by
David West Books

Editor: Gail Bushnell

Photo credits: All photographs on pages 4–7 and 44–45 by NASA.

Library of Congress Cataloging-in-Publication Data

West, David, 1956-
   Astronauts / by David West; illustrated by Jim Robins.
      p. cm. -- (Graphic careers)
   Includes index.
   ISBN 978-1-4042-1461-3 (library binding) -- ISBN 978-1-4042-1462-0
(pbk.) -- ISBN 978-1-4042-1463-7 (6 pack)
   1. Astronautics--Vocational guidance--Juvenile literature. 2.
Astronauts--Juvenile literature. I. Robbins, Jim, 1949- ill. II.
Title.
   TL793.W469 2007
   629.450023--dc22

                                      2007045208

Manufactured in China

CONTENTS

LIVING AND WORKING IN SPACE

Manned space stations have been orbiting the Earth since 1971. The early Russian space station program, called Salyut, carried out research into living in space, and paved the way for the later multimodular space stations, such as Mir and the International Space Station (ISS).

Astronauts are ferried to the ISS either by the Space Shuttle (below), or a Salyut spacecraft (above).

LIVING AND WORKING IN ZERO GRAVITY

Astronauts on board the ISS continue to research the effects of long-term space flight on the human body. They also carry out scientific studies and experiments.

The duration record of 437.7 days was set by Valeriy Polyakov aboard Mir (above) from 1994 to 1995. The station existed until March 23, 2001, at which time it was deliberately deorbited, breaking apart during atmospheric reentry over the South Pacific Ocean.

Living and working in space brings a unique set of problems. Regular exercise is a must to keep muscles and bones from degenerating (1). Because there is no gravity to keep things on the floor or tabletop, everything has to be strapped or Velcroed down. This includes astronauts when they are sleeping (2). Food crumbs and moisture must not get into the electronics, so care has to be taken with the specially prepared food and drink packs (3).

The ISS is continually being added to, which requires astronauts to work in space wearing space suits.

The ISS is in low Earth orbit and can be seen from Earth by the naked eye. It orbits approximately 205 miles (330 kilometers) above the Earth. It has been continually habited since November 2, 2000, and so far has been visited by astronauts from fourteen countries.

ASTRONAUT TRAINING

Astronauts may be a commander/pilot, mission specialist, or payload specialist.

ASTRONAUTS

Commander/pilots fly the Shuttle and may deploy satellites using the Shuttle's mechanical arm. Training includes flying Shuttle simulators and Shuttle training aircraft. Mission specialists have a specific job to do, such as a spacewalk to attach a new part to the ISS. Training requires hundreds of hours in the Neutral Buoyancy Laboratory (NBL), a large water tank that simulates zero gravity, and on computer simulators and virtual reality programs. The payload specialist performs any specialized duties the mission requires. Payload specialists are people other than NASA personnel, and some are foreign nationals.

1. Mission specialist using virtual reality hardware in the Space Vehicle Mockup Facility. 2. Testing an Extravehicular Mobility Unit (EMU) space suit on board a KC-135 aircraft, which provides zero gravity for twenty seconds at a time. 3. An astronaut at the Russian space program on a Soyuz simulator. 4. The Space Shuttle cockpit simulator. 5. Astronauts practice landings in the KC-135 as it is similar in gross weight to a Shuttle. 6. Training on the motion-based Shuttle simulator. 7. Canadian D. Williams during mission specialist training in the NBL.

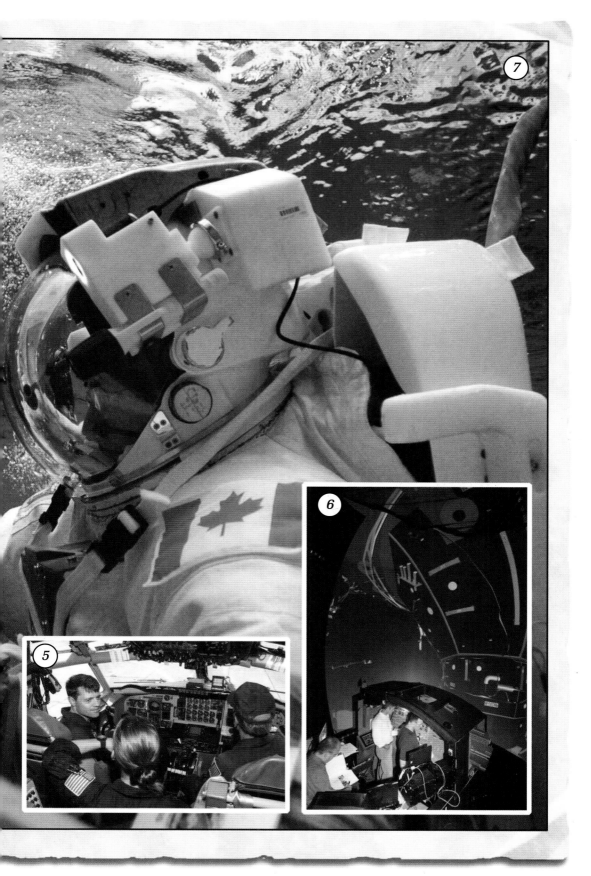

YURI ALEXEYEVICH GAGARIN
COSMONAUT
THE FIRST PERSON IN SPACE

APRIL 12, 1961, SOUTHWEST OF ENGELS SMELOVKA, SARATOV, USSR*. A YOUNG GIRL RUNS SCREAMING TO HER MOTHER, WHO WATCHES AN APPROACHING FIGURE IN STRANGE CLOTHING.

DON'T BE AFRAID. I AM A SOVIET, A FRIEND.

I MUST FIND A TELEPHONE TO CALL MOSCOW.

*UNION OF SOVIET SOCIALIST REPUBLICS.

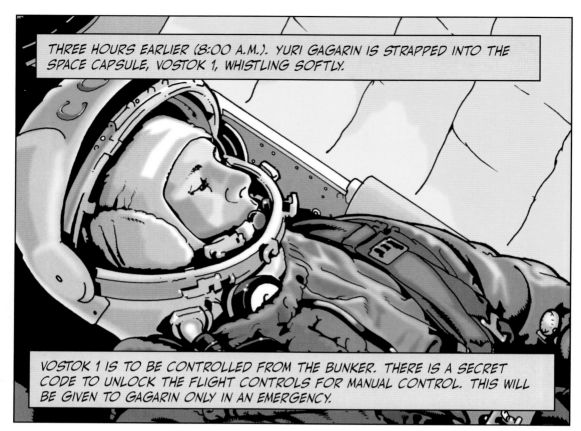

THREE HOURS EARLIER (8:00 A.M.). YURI GAGARIN IS STRAPPED INTO THE SPACE CAPSULE, VOSTOK 1, WHISTLING SOFTLY.

VOSTOK 1 IS TO BE CONTROLLED FROM THE BUNKER. THERE IS A SECRET CODE TO UNLOCK THE FLIGHT CONTROLS FOR MANUAL CONTROL. THIS WILL BE GIVEN TO GAGARIN ONLY IN AN EMERGENCY.

IT IS TIME FOR IVANOVSKY TO CLOSE AND SEAL THE HATCH. HE FEELS IT IS WRONG TO KEEP THE CODE FROM GAGARIN.

YURI, THE CODE NUMBERS ARE...

YES, I KNOW. KOROLEV TOLD ME, AND KAMININ. THANK YOU ANYWAY

THE HATCH IS REATTACHED WITH NO PROBLEMS. KOROLEV GIVES THE ORDER FOR LAUNCH.

IGNITION IS BEING GIVEN, KEDR. I AM ZARYA-1.

AT 9:07 A.M. MOSCOW TIME, THE R-7 ROCKET CARRYING VOSTOK 1 LIFTS OFF.

COMPLETE TAKEOFF.

LET'S GO!

20 SECONDS INTO THE FLIGHT.

I AM ZARYA-1. WE ARE ALL WISHING YOU A GOOD FLIGHT. IS EVERYTHING OK?

THANK YOU. BYE-BYE. SEE YOU SOON, DEAR FRIENDS.

TWO MINUTES INTO THE FLIGHT, THE ROCKET BOOSTERS FALL AWAY.

THREE MINUTES INTO THE FLIGHT THE NOSE CONE FALLS AWAY, AS EXPECTED.

KEDR. I CAN SEE THE EARTH. VISIBILITY IS FINE.

THE PRESSURE OF G-FORCES BEGINS TO PRESS GAGARIN BACK INTO HIS SEAT.

I AM ZARYA-1. HOW ARE YOU FEELING? I AM ZARYA—OVER.

I AM KEDR. I AM FEELING FINE.

IN THE BUNKER, THE SIGNALS RETURNING FROM THE ROCKET ENGINES ARE BAD. THEY SHOULD BE IN "FIVES" BUT THEY ARE READING IN "THREES."

IF THEY GET TO "TWOS" IT WILL BE A DISASTER.

EVENTUALLY THEY RETURN TO "FIVES." KOROLEV IS ALMOST SICK WITH RELIEF.

I AM ZARYA-1. ALL IS WELL. EVERYTHING IS WORKING.

I HEAR YOU. I CAN FEEL IT WORKING. I AM WATCHING THE EARTH.

ABOUT TEN MINUTES INTO THE FLIGHT, THE UPPER-STAGE ENGINES DROP AWAY. VOSTOK 1 IS TRAVELING AT 28,000 MILES (45,000 KILOMETERS) PER HOUR. IT HAS REACHED SPACE AND GAGARIN, THE FIRST HUMAN IN SPACE, BEGINS TO EXPERIENCE WEIGHTLESSNESS.

THE FEELING OF WEIGHTLESSNESS IS INTERESTING. EVERYTHING IS FLOATING. BEAUTIFUL. INTERESTING.

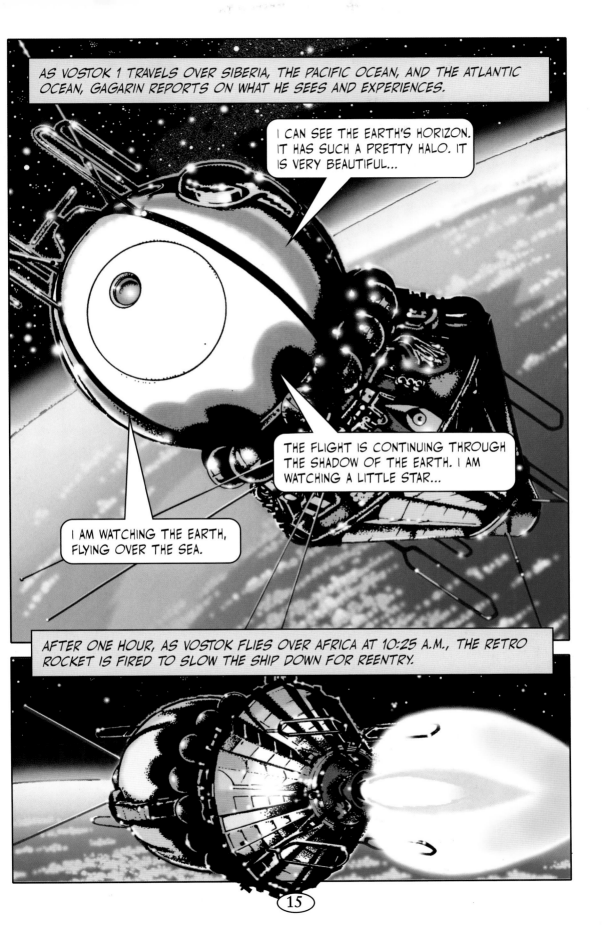

SUDDENLY, THERE IS A SICKENING JOLT. VOSTOK STARTS SPINNNG.

THIS IS NOT RIGHT.

THE INSTRUMENT MODULE, INSTEAD OF SEPARATING FROM THE DESCENT CAPSULE, IS STILL ATTACHED BY SOME CABLES.

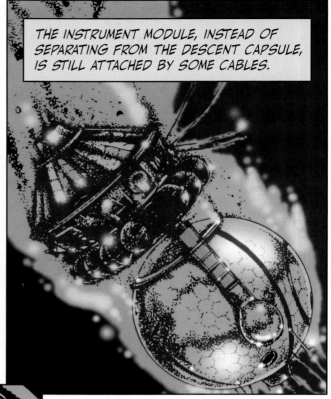

AS THE TWO PARTS SPIN AROUND EACH OTHER, THE G-FORCES THREATEN TO CAUSE GAGARIN TO BLACK OUT.

GNNN. MUST... STAY... CONSCIOUS.

THE CAPSULE GETS HOTTER AND HOTTER AS THE FRICTION OF THE AIR OVER THE CAPSULE BURNS THE OUTER LAYERS.

AT 10:35 A.M. THE CABLES HOLDING THE TWO PARTS BURN AWAY AND THEY SEPARATE. THERE ARE STILL 20 MINUTES OF DESCENT TO GO AS GAGARIN ENDURES EIGHT G'S.

AT 10:55 A.M. THERE IS A LOUD EXPLOSION AS THE ESCAPE HATCH IS EJECTED AT FOUR AND A HALF MILES (SEVEN KILOMETERS) ABOVE THE EARTH.

BLAM

SECONDS LATER, GAGARIN IS BLASTED FROM THE CAPSULE IN HIS EJECTOR SEAT.

LOOK, MAMA. WHAT IS IT?

AS HE BEGINS TO FALL, HIS PARACHUTE OPENS AND HIS SEAT FALLS AWAY. AT THE SAME TIME THE CAPSULE'S PARACHUTE OPENS AND BOTH MAN AND MACHINE FLOAT DOWN TO EARTH.

THE END

Dr. Jerry Linenger
ASTRONAUT
Fire in Space

FEBRUARY 24, 1997. AN AMERICAN ASTRONAUT JERRY LINENGER HAS BEEN ON THE RUSSIAN SPACE STATION, MIR, FOR FIVE WEEKS WITH COSMONAUTS VALERI KORZUN AND ALEXANDER KALERI. TWO WEEKS EARLIER, A RUSSIAN REPLACEMENT TEAM OF VASILY TSIBLIEV, ALEXANDER LAZUTKIN, AND GERMAN RESEARCHER REINHOLD EWALD HAD ARRIVED IN A SOYUZ CAPSULE.

MMM. JELLIED PIKE PERCH AND BORSCHT IS MY FAVORITE SPACE FOOD.

EVERYTHING IS YOUR FAVORITE SPACE FOOD.

THERE WAS A TWENTY-DAY OVERLAP BEFORE THE OLD CREW RETURNED TO EARTH. THE SPACE STATION WAS CRAMPED WITH THE SIX OF THEM, AND IT PUT A STRAIN ON THE OXYGEN SUPPLY.

LINENGER LOOKS DOWN THE TUNNEL TOWARD THE KVANT 1 MODULE.

NOT GOOD.

A FLAME SHOOTS OUT THREE FEET (ONE METER) FROM THE OXYGEN CANISTER.

THE FIRE IS SO HOT IT'S MELTING METAL.

BE PREPARED FOR DECOMPRESSION!

SMOKE RAPIDLY FILLS THE SPACE STATION. HOLDING HIS BREATH, LINENGER HEADS BACK TO THE SPEKTR MODULE TO FIND AN OXYGEN RESPIRATOR.

NOT GOOD.

THE FIRE COULD BURN A HOLE IN THE SKIN OF THE SPACE STATION.

FINDING ONE, HE RIPS IT OFF THE WALL, QUICKLY PUTS THE MASK OVER HIS FACE, AND SWITCHES ON THE AIRFLOW.

IT DOESN'T WORK!

DESPERATE FOR AIR, LINENGER GROPES HIS WAY THROUGH THE THICK SMOKE.

OPEN A WINDOW! NO, THAT'S ABSURD!

HE FEELS THE BULGE OF A RESPIRATOR CASE. RIPPING IT OPEN, HE PULLS THE MASK TO HIS FACE AND SWITCHES ON THE AIR.

AHHHHHH, I CAN BREATHE.

AFTER A MINUTE HE STRAPS THE MASK ON PROPERLY AND MEETS THE OTHERS AT THE MAIN NODE.

LINENGER AND KORZUN HEAD TOWARD THE FIRE WITH A FIRE EXTINGUISHER, WHILE THE OTHERS PREPARE A SOYUZ CAPSULE FOR EVACUATION.

THERE'S ONLY ROOM FOR THREE PEOPLE IN THE SOYUZ. WE MUST PUT OUT THIS FIRE.

LINENGER HOLDS ON TO KORZUN WHILE HE USES THE FIRE EXTINGUISHER.

IT'S NOT MAKING ANY DIFFERENCE.

AFTER USING THREE FIRE EXTINGUISHERS, THE FIRE TAKES FOURTEEN MINUTES TO BURN OUT.

THE FIRE IS OUT!

THE SMOKE LINGERS IN THE STATION FOR SOME TIME. IT TAKES A WHILE BEFORE IT BECOMES SAFE TO BREATHE THE AIR. LINENGER TREATS SOME CREW MEMBERS WITH MINOR BURNS, BUT THERE ARE NO SERIOUS INJURIES. LINENGER STAYED ON MIR FOR 132 DAYS AND WAS PICKED UP BY ATLANTIS IN MAY, 1997.   **THE END**

Thomas D. Jones
ASTRONAUT
Space Shuttle Mission STS-98

AUGUST 2000. THE CREW OF STS-98 ARE: COMMANDER KENNETH D. "TACO" COCKRELL, PILOT MARK L. "ROMAN" POLANSKI, AND MISSION SPECIALISTS ROBERT L. "BEAMER" CURBEAM, MARSHA S. IVINS, AND THOMAS D. JONES.

TWO IMUS* ARE DOWN! OH, HECK, THERE GOES THE LAST ONE!

FIFTY THOUSAND FEET, ONE POINT TWO G'S.

*INERTIAL MEASURING UNITS. WITHOUT AT LEAST ONE OF THESE OPERATING, THE SHUTTLE ORBITER CANNOT CONTROL ITS GLIDE BACK TO EARTH.

AS MISSION CONTROL ADVISES THE CREW, THE SHUTTLE COMMANDER, KENNETH "TACO" COCKRELL, HITS THE BFS* BUTTON.

ATLANTIS, ENGAGE BFS.

*BACKUP FLIGHT SYSTEM.

EVERYBODY HEAVES A SIGH OF RELIEF AS THE THIRD IMU STARTS TO OPERATE AGAIN.

PHEW!

SUDDENLY, THE ORBITER ROLLS VIOLENTLY...

WHOA!

WE'RE ALL DEAD...

26

AS MISSION SPECIALISTS, ROBERT CURBEAM AND TOM SPEND HUNDREDS OF HOURS IN THE NEUTRAL BUOYANCY LABORATORY (NBL). THEY ARE TRAINING FOR THE THREE SPACE WALKS WHEN THEY WILL ATTACH THE NEW DESTINY LAB TO THE ISS.

THE SPACE WALKS WILL LAST SIX HOURS EACH, SO THERE IS PLENTY OF TRAINING TO GET THROUGH. IT IS HARD, DEMANDING WORK.

NOT ALL THE TRAINING IS HARD WORK. ROBERT AND TOM HAVE CLASSES IN THE VIRTUAL REALITY LAB. HERE, THEY PUT ON A PAIR OF 3-D TELEVISION GOGGLES AND PRACTICE USING THE NITROGEN-POWERED JET PACK. THIS IS IN CASE THE TETHERS BREAK AND THEY FLOAT OFF INTO SPACE.

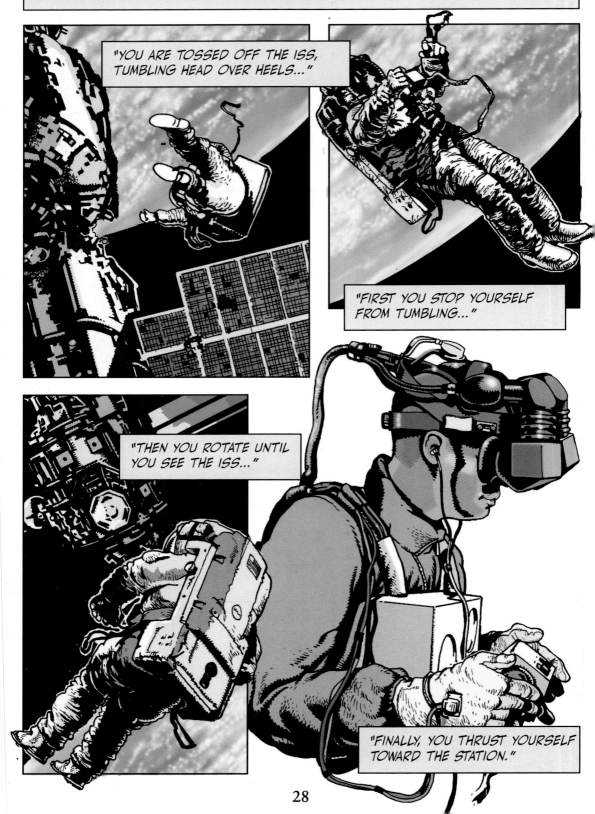

"YOU ARE TOSSED OFF THE ISS, TUMBLING HEAD OVER HEELS..."

"FIRST YOU STOP YOURSELF FROM TUMBLING..."

"THEN YOU ROTATE UNTIL YOU SEE THE ISS..."

"FINALLY, YOU THRUST YOURSELF TOWARD THE STATION."

WHILE THE CREW OF STS-98 CONTINUES TRAINING, THE FIRST CREW TO TAKE UP RESIDENCE IN THE ISS ARRIVES AT THE BEGINNING OF NOVEMBER 2000. THEY ARE BILL SHEPHERD, SERGEI KRIKALEV, AND YURI GIDZENKO.

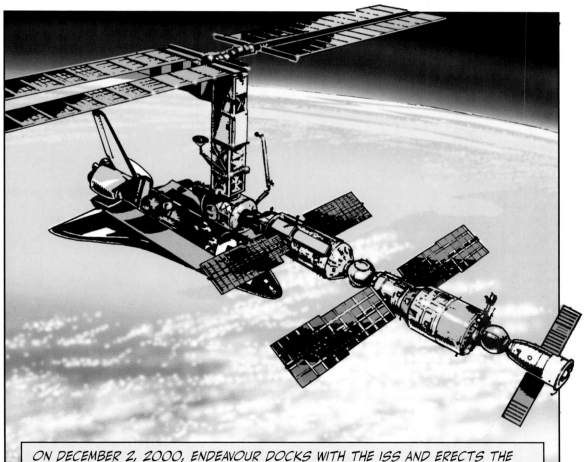

ON DECEMBER 2, 2000, ENDEAVOUR DOCKS WITH THE ISS AND ERECTS THE P6 TRUSS WITH A PAIR OF SOLAR ARRAYS. STS-98'S TURN IS NEXT...

AFTER LONG DELAYS, THE DATE FOR LAUNCH IS FEBRUARY 7, 2001, AT KENNEDY SPACE CENTER (KSC). THE CREW ENTERS QUARANTINE ON JANUARY 31. THIS KEEPS THEM AWAY FROM DAY-TO-DAY ENCOUNTERS WITH COLLEAGUES AND FAMILY THAT MIGHT EXPOSE THEM TO INFECTION.

ON FEBRUARY 4, THE CREW FLIES TO KSC, WHERE THEIR FAMILIES ARE WAITING.

THE CREW AND THEIR FAMILIES ARE ALLOWED TO LOOK OVER ATLANTIS.

THREE DAYS LATER, THE CREW MEMBERS ARE ALL STRAPPED IN, READY FOR TAKEOFF. TOM IS ALONE ON THE MID DECK. ON THE RETURN TRIP, HE WILL SIT ON THE UPPER DECK WITH THE REST OF THE CREW.

OK, TOM, YOU'RE GOOD TO GO.

THANKS, RAY J, I'LL SEE YOU IN A COUPLE OF WEEKS.

GREG "RAY J" JOHNSON AND THE REST OF THE "CLOSE-OUT CREW" LEAVE THE CABIN AND CLOSE THE ORBITER HATCH.

TIME TO TAKE MY ANTINAUSEA PILL.

THE CALM VOICES OF THE LAUNCH TEAM GO THROUGH THE COUNTDOWN LIST.

T-MINUS-SIXTY MINUTES...

5:53 P.M.

WE ARE GOOD TO PROCEED.

COPY THAT. THE CREW THANKS THE KSC TEAM. LET'S GET THIS VEHICLE OFF THE LAUNCH PAD.

T-MINUS-NINE MINUTES AND COUNTING. GLS* HAS BEEN INITIATED.

*GROUND LAUNCH SEQUENCER.

THE GANTRY SWING ARM RETRACTS.

PLT*, PERFORM APU* START-UP.

*PLT = SHUTTLE PILOT.
*APU = AUXILIARY POWER UNITS.

APU START COMPLETE.

THREE MINUTES AND COUNTING.

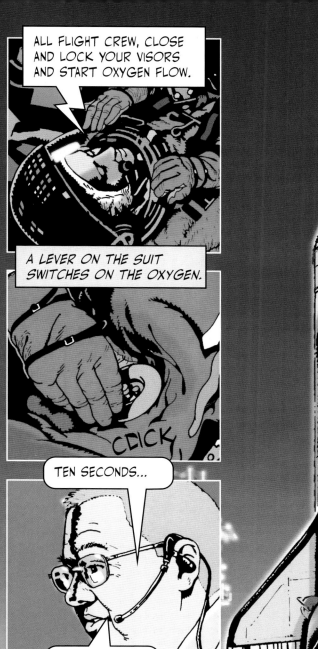

ALL FLIGHT CREW, CLOSE AND LOCK YOUR VISORS AND START OXYGEN FLOW.

A LEVER ON THE SUIT SWITCHES ON THE OXYGEN.

CLICK

TEN SECONDS...

GO FOR MAIN ENGINE START.

RUMBLE

BOOM

JUST AFTER THE THREE MAIN ENGINES FIRE, THE TWO SOLID ROCKET BOOSTERS (SRBS) FIRE.

AS THE SHUTTLE ACCELERATES, THE CREW IS SUBJECTED TO TWO AND A HALF G'S. THE ACCELERATION LESSENS AS THE TWO SRBS RUN OUT OF FUEL.

SEVEN MINUTES INTO THE FLIGHT, AND ATLANTIS IS TRAVELING AT NINETEEN TIMES THE SPEED OF SOUND, SUBJECTING THE CREW TO OVER THREE G'S.

HEY, THERE'S A GORILLA SITTING ON MY CHEST!

EXPLOSIVE BOLTS SHEAR THE TWO SRBS FROM THE EXTERNAL FUEL TANK.

PARACHUTES SLOW THEIR DESCENT TO THE OCEAN, WHERE THEY WILL BE PICKED UP TO BE REUSED ON A FUTURE SHUTTLE MISSION.

AT EIGHT MINUTES INTO THE FLIGHT, THE MAIN ENGINES ARE CUT OFF, AND ATLANTIS ROLLS 180 DEGREES. TEN SECONDS LATER, THE EXTERNAL FUEL TANK IS RELEASED WITH A CLANG.

THE PRIMARY RCS* JETS NUDGE THE ORBITER AWAY FROM THE EMPTY TANK.

*REACTION CONTROL SYSTEM.

EVENTUALLY THE TANK WILL BURN UP AS IT ENTERS THE EARTH'S ATMOSPHERE.

GOOD JOB, PEOPLE.

FLOATING AROUND IN ZERO GRAVITY, THE CREW CHANGES OUT OF LAUNCH SUITS AND STARTS PREPARING ATLANTIS'S CABIN FOR LIVING IN.

I'M GLAD I TOOK MY ANTINAUSEA PILLS.

FACES PUFF UP AS FLUID, NO LONGER PULLED DOWN BY GRAVITY, MOVES TO HEAD AND CHEST. THE INNER EAR, USED FOR BALANCE, IS ALSO AFFECTED, LEADING TO MOTION SICKNESS.

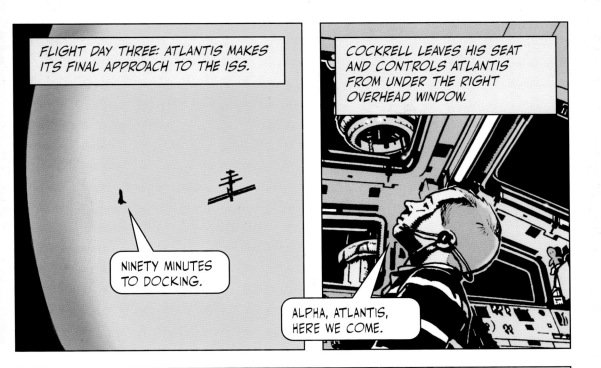

FLIGHT DAY THREE: ATLANTIS MAKES ITS FINAL APPROACH TO THE ISS.

NINETY MINUTES TO DOCKING.

COCKRELL LEAVES HIS SEAT AND CONTROLS ATLANTIS FROM UNDER THE RIGHT OVERHEAD WINDOW.

ALPHA, ATLANTIS, HERE WE COME.

TRAVELING AT FIVE MILES PER SECOND, THE SLIGHTEST WRONG MOVE COULD CAUSE DISASTER. (IN 1997, RUSSIA'S MIR SPACE STATION WAS HIT BY A CARGO SPACESHIP, PUNCTURING A HOLE IN MIR'S HULL.)

EIGHT FEET, POINT ONE ZERO... SIX FEET POINT ONE...

CAPTURE!

MARSHA MAKES THE RANGE CALLS.

THE ORBITER SHAKES AS THE DOCKING RINGS SLAM TOGETHER

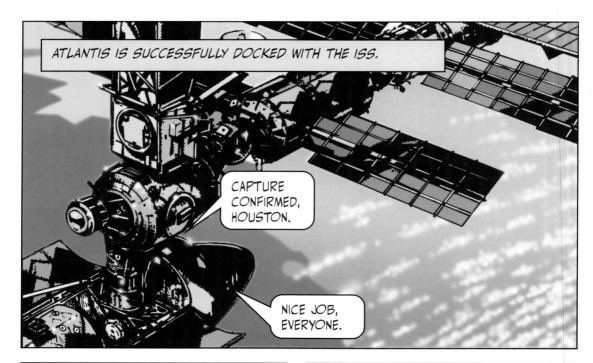

ATLANTIS IS SUCCESSFULLY DOCKED WITH THE ISS.

CAPTURE CONFIRMED, HOUSTON.

NICE JOB, EVERYONE.

ROBERT EQUALIZES THE PRESSURE IN THE JOINING TUNNEL FROM THE ATLANTIS'S AIRLOCK.

OK, TOM, LET'S SEE HOW YOU DO WITH THIS HATCH.

HA, HA. I SHOULD HAVE SEEN THAT ONE COMING.

ON HIS PREVIOUS MISSION, TOM AND FELLOW ASTRONAUT TAMMY JERNIGAN HAD TO ABORT THEIR FIRST SPACE WALK WHEN THE AIRLOCK HATCH MALFUNCTIONED.

IT'S NOT BUDGING.

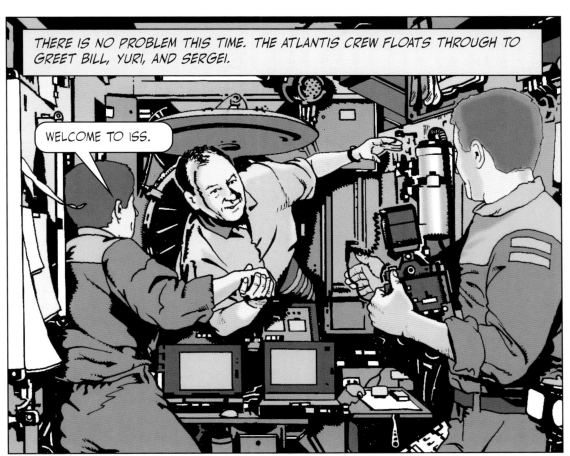

THERE IS NO PROBLEM THIS TIME. THE ATLANTIS CREW FLOATS THROUGH TO GREET BILL, YURI, AND SERGEI.

WELCOME TO ISS.

AFTER A BRIEF TOUR THE CREWS RETURN TO ATLANTIS, WHERE SEVERAL BAGS OF CARGO ARE HANDED OVER.

FRESH FOOD!

FOUR HOURS AFTER DOCKING, ATLANTIS'S CREW PREPARES FOR THE EVA* THE NEXT DAY. TOM AND ROBERT WEAR MASKS ATTACHED TO PORTABLE OXYGEN BOTTLES TO GET RID OF NITROGEN FROM THEIR BLOOD. THIS WILL STOP THEM FROM GETTING THE BENDS.

*EXTRA-VEHICULAR ACTIVITY.

39

ON FEBRUARY 10, ROBERT AND TOM ARE WAITING IN THE AIRLOCK, READY FOR THEIR SPACE WALK. THEY ARE NOW BREATHING PURE OXYGEN TO FLUSH THE LAST OF THE NITROGEN FROM THEIR BLOOD.

CHECKLIST COMPLETE. AIRLOCK DEPRESS VALVE READY.

WHOQOOSH

DEPRESS VALVE TO FIVE.*

*THE AIRLOCK IS DEPRESSURIZED TO 5 PSI, TO CHECK THE SUITS FOR ANY LEAKS.

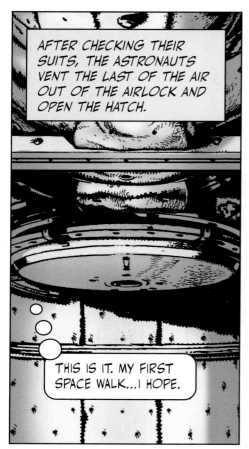

AFTER CHECKING THEIR SUITS, THE ASTRONAUTS VENT THE LAST OF THE AIR OUT OF THE AIRLOCK AND OPEN THE HATCH.

THIS IS IT. MY FIRST SPACE WALK...I HOPE.

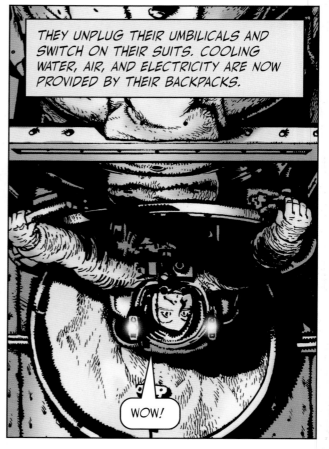

THEY UNPLUG THEIR UMBILICALS AND SWITCH ON THEIR SUITS. COOLING WATER, AIR, AND ELECTRICITY ARE NOW PROVIDED BY THEIR BACKPACKS.

WOW!

TOM HOOKS HIMSELF AND THEN ROBERT ONTO THEIR 50-FOOT (15-METER) SAFETY LINES.

THE ASTRONAUTS QUICKLY START THEIR WORK. THE $1.4-BILLION DESTINY MODULE HAS TO BE INSTALLED WITHIN THIRTY-SIX HOURS, OTHERWISE THE COLD OF SPACE WILL FREEZE THE COOLANT, BURSTING THE HEAT EXCHANGERS.

OK, BEAMER, WE'RE READY TO GO.

IF I LET GO, I FEEL I'LL FALL TWO HUNDRED MILES TO EARTH.

HEH, HEH. IT'S ONLY AN ILLUSION. YOU'LL GET USED TO IT.

IT'S LIKE BALANCING ON A UNICYCLE WHILE TRYING TO SPIN A LUG NUT OFF A WHEEL!

HOW ARE YOU DOING, TOM?

YIKES! THESE BOLTS ARE FROZEN STIFF.

LATER, AS TOM TRIES TO UNDO SOME BOLTS, THE POWER TOOL IS TWISTED OUT OF HIS HANDS. IT TAKES A WHILE TO FREE THE BOLTS.

41

THE VALVES WERE FIXED, AND EVENTUALLY, AFTER SIX HOURS OF EVA, THE ASTRONAUTS WERE READY TO HEAD BACK TO THE AIRLOCK. BEFORE THEY COULD ENTER, THOUGH, TOM HAD TO BRUSH OFF THE AMMONIA FROST-DUST THAT HAD SETTLED ON ROBERT'S SUIT.

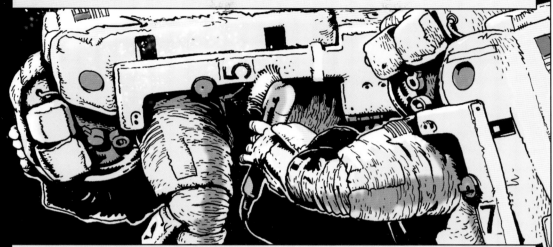

AFTER SIX AND A HALF HOURS OF EVA, TOM AND ROBERT MADE IT BACK INSIDE ATLANTIS. DESPITE THE DRAMA, TOM'S FIRST SPACE WALK HAD BEEN A SUCCESS. THE DESTINY LAB WAS POWERED UP AND THE COOLING SYSTEM WORKED, HAVING LOST ONLY FIVE PERCENT OF ITS COOLANT. THE FOLLOWING DAY, THE LAB WAS FITTED OUT AND READY FOR BUSINESS.

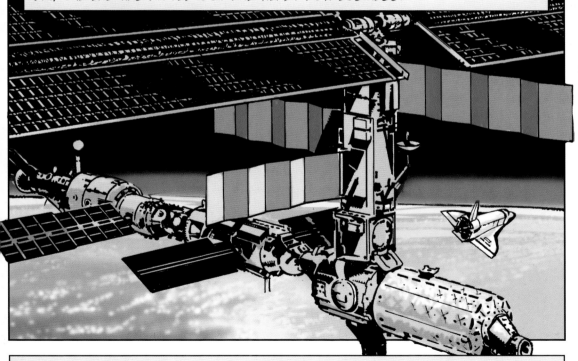

TOM AND ROBERT PERFORMED TWO MORE SPACE WALKS, WHICH INCLUDED THE 100TH SPACE WALK BY U.S. ASTRONAUTS. THE MISSION WAS A SUCCESS AND ATLANTIS RETURNED TO EARTH, TOUCHING DOWN IN THE AFTERNOON OF FEBRUARY 7, 2001, TWELVE DAYS, TWENTY-ONE HOURS AFTER LAUNCH.   **THE END**

HOW TO BECOME AN ASTRONAUT

NASA–with its international partners Japan, Canada, Russia, and the European Space Agency (ESA)–accepts applications for the Astronaut Candidate Program on a continual basis.

BASIC QUALIFICATIONS

To be a pilot or mission specialist, you have to be a U.S. citizen. You will need a bachelor's degree in engineering, biological sciences, physical sciences, or mathematics from an accredited college or university, as well as three years of related experience. (A master's degree equals one year of experience, and a doctorate equals three years.) Pilots need to pass a Class I physical exam. Mission/payload specialists must pass a Class II physical. Pilots need more than 1,000 hours experience as pilot in command of a jet aircraft. Height requirements are 64–76 inches (163–193 cm) for pilots, 58.5–76 inches (149–193 cm) for mission/payload specialists. After you fill out an application form, NASA screens the application, and you may be asked to go to a week-long session to participate in interviews, medical tests, and orientations. Your performance will be evaluated, and if you are lucky, you may be accepted as an astronaut candidate. NASA announces candidates every two years, selecting about a hundred men and women out of thousands of applicants.

Astronaut candidates tumble in zero gravity inside a KC-135. They will spend two years training before being selected as an astronaut for a specific mission in space.

GLOSSARY

ammonia A colourless, pungent gas.

antinausea pills Pills that prevent a person from feeling sick.

bends A painful sickness that occurs when nitrogen in the blood expands due to a sudden reduction in air pressure. The bends can be fatal.

cosmonaut A Russian astronaut.

decompression A reduction in air pressure.

deorbit To slow down a spacecraft in orbit so that it falls to Earth.

ejector seat A rocket-propelled seat that carries an astronaut clear of his spacecraft.

g (pronounced jee) G stands for gravity and is short for g-force.

g-force The force a person feels when accelerating. One g is equal to the force acting on a body while standing on Earth.

Kvant 1 module The first addition to the Mir base block which contained scientific instruments for observations and experiments.

multimodular Having many modules.

node Connecting module on a space station.

P6 truss Structure that supports the P6 solar arrays.

quarantine To place a person in an area away from other people so that he or she does not catch an infection.

respirator Self-contained breathing apparatus.

retro rockets Rockets that slow down a spacecraft by firing in the direction of travel.

scrubbers An apparatus that is used for removing impurities from a gas.

simulator A machine with a similar set of controls to the real machine that allows people to train safely.

slurry A runny mixture of solid particles and water.

Spektr module The fifth module of the Mir Space Station. It was designed for remote observation of Earth's environment, containing atmospheric and surface research equipment.

tethers Safety lines that attach an astronaut to a spacecraft.

umbilicals Tubes that carry oxygen and electricity from a spacecraft to an astronaut's space suit.

unique Unlike anything else.

virtual reality A computer-simulated reality that can be interacted with.

zero gravity Weightlessness.

FOR MORE INFORMATION

ORGANIZATIONS

Public Communications and Inquiries Management Office
NASA Headquarters
Suite 5K39
Washington, D.C. 20546-0001
(202) 358-0001
Web site: http://www.nasa.gov

John F. Kennedy Space Center
Cape Canaveral, FL 32899
Web site: http://www.kennedyspacecenter.com

FOR FURTHER READING

Cole, Michael, D. *Astronauts: Training for Space* (Countdown to Space). Berkeley Heights, NJ: Enslow Publishers, 1999.

Hansen, Ole Steen. *The Story of Flight: Space Flight.* New York, NY: Crabtree Publishing, 2004.

Hibbert, Clare. *The Inside & Out Guide to Spacecraft.* Chicago, IL: Raintree, 2006.

Jeffrey, Gary. *Graphic Discoveries: Incredible Space Missions.* New York, NY: Rosen Publishing, 2008.

Jones, Tom. *Sky Walking: An Astronaut's Memoir.* New York, NY: HarperCollins, 2006.

Space Heroes: Amazing Astronauts. London, England: Dorling Kindersley, 2004.

INDEX

Web Sites

Due to the changing nature of Internet links, Rosen Publishing has developed an online list of Web sites related to the subject of this book. This site is updated regularly. Please use this link to access the list:

http://www.rosenlinks.com/gc/astr